Cláudia Leite Munhoz

Extraction of pectin from guava cultivar Pedro Sato

Cláudia Leite Munhoz

Extraction of pectin from guava cultivar Pedro Sato

Effect of extraction conditions

ScienciaScripts

Imprint

Cover image: www.ingimage.com

This book is a translation from the original published under ISBN 978-620-2-03229-2.

Publisher:
Sciencia Scripts
is a trademark of
Dodo Books Indian Ocean Ltd. and OmniScriptum S.R.L publishing group

120 High Road, East Finchley, London, N2 9ED, United Kingdom
Str. Armeneasca 28/1, office 1, Chisinau MD-2012, Republic of Moldova, Europe
Managing Directors: Ieva Konstantinova, Victoria Ursu
info@omniscriptum.com

Printed at: see last page
ISBN: 978-620-6-74802-1

To God for his presence in my life.

To my parents, José and Raquel, for their love, support and trust.

ACKNOWLEDGEMENTS

To everyone who has helped me in this stage of my life, in particular:

To God, for his undeniable presence in my life.

To my family, for their support and affection at all times.

To Prof Dr Manoel Soares Soares Júnior, for his guidance and support.

To Prof Dr Eliana Janet Sanjinez-Argandona, for her guidance and, above all, for her friendship.

The Federal University of Goiás, for the opportunity to study for a master's degree.

To CAPES for the master's scholarship.

The Federal University of Grande Dourados, for making its laboratories available to carry out the research.

To the professors of the Postgraduate Programme at UFG, for the subjects taught, especially Prof. Dr. [(a)]. Dr Maria Margareth Veloso Naves for her affection and dedication.

To my classmates for their encouragement, support and friendship during this period, especially Letícia Fleury Viana for the friendship that emerged.

To Ayd Mary Oshiro, for her support, friendship and companionship during the guava research.

To the employees of the State Animal Health Defence Agency in Vicentina-MS, for taking time off work due to the research.

"How can a young man keep his way pure? By observing it ***according to your word. "***

Psalm 119. 9

SUMMARY

CHAPTER 1

INTRODUCTION

Pectin is a polysaccharide found in the cell wall of higher plants and is responsible for the structure of fruit. The molecule is basically made up of D-galacturonic acid linked by alpha-(1,4) bonds. Pectins from different plants have a common chemical structure, but there may be differences in their composition and structure during the ripening phase.

Pectins can be classified into high and low degrees of esterification, depending on the amount of esterified galacturonic acid. Highly esterified pectins (over 50%) form gels after heating in solutions with a sugar concentration of over 60% and a pH of less than 3.6. Low esterification pectins (below 50%) form gels in the presence of bivalent ions such as calcium.

The application of pectin in the food industry is very versatile, as this additive has gelling, thickening and stabilising properties and can be used in a wide variety of dairy, meat and bakery products, jams, fruit pulp, drinks and more.

Generally speaking, the process of obtaining pectin includes the following steps: aqueous extraction of the material used, separation of the pectin from the liquid extract and drying. Extraction methods can be carried out using acids, bases or enzymes, with the acid process being the most common. Separation of the pectin from the extract usually takes place in the presence of alcohols and drying is carried out at mild temperatures between 40 and 60 °C. The pectins extracted from fruit are generally highly esterified, although there may be some that are less esterified. The extraction method has a direct influence on the yield and properties of the pectin. Inorganic acids such as hydrochloric and nitric are used in commercial extraction, which generates toxic waste. With the growing demand for alternative processes in which non-toxic waste can be obtained and utilised, there is a trend towards using organic acids such as citric acid.

There have been several studies on how to utilise different waste products to extract pectin, however, despite guava (Psidium guajava L.) containing a reasonable amount of pectin, there is little information on its use as a source of extraction. In addition, there is a great deal of post-harvest loss of this fruit. Pectin extraction studies using guava could help to utilise the fruit, thus reducing losses.

In view of the above, the objectives of this work were to characterise the raw material in natura, the flours of the guava cv Pedro Sato fractions and to optimise the factors that influence the yields of pectin extraction from the flours of guava pulp and pulp with peel. As well as characterising the pectins obtained at the optimum points in terms of their degree of esterification, comparing them with commercial pectin.

CHAPTER 2

LITERATURE REVIEW

2.1 PECTIN

There is great interest in research into pectin because of its gelling, stabilising and thickening properties in foods, because it is a dietary fibre and because it is an important component of plant cell walls. Due to evidence that pectin reduces plasma cholesterol levels and regulates blood glucose, there is a growing increase in the consumption of foods and food products with high levels of this soluble fibre (KLIEMANN, 2006).

2.1.1 Features

Pectin is a polysaccharide that makes up the cell wall of dicotyledonous plants. Its function is to contribute to adhesion between cells and to the mechanical resistance of the cell wall. The association of pectin with cellulose and hemicellulose gives rise to protopectin in plant tissues. Protopectin, which is insoluble, is easily hydrolysed by heating in an acidic environment, forming pectin (ORDONEZ-PEREDA, 2005; THAKUR; SINGH; HANDA, 1997).

In immature plant tissues, protopectin predominates and as the fruit matures, calcium is released and the pectic polymer is solubilised by the action of specific enzymes (pectinamethylesterase and polygalacturonase), forming pectin. The enzyme pectinamethylesterase (PME) is responsible for breaking the methyl ester bonds and the enzyme polygalacturonase (PG) transforms the galacturonic acid polymers into pectic acids, which are soluble in water. The process of pectin solubilisation contributes to the softening of tissues by reducing the cohesive force between cells due to enzymatic action (DA-SILVA; FRANCO; GOMES, 1997). The physiological changes in fruit during ripening, storage or processing are related to the transformation of pectins in terms of their degree of esterification, molecular weight and degree of acetylation (MAY, 1990).

The chemical structure of pectin (Figure 1) has been the subject of scientific research. Elucidating the structure of the pectin molecule is important for understanding its function in plant growth and development, during fruit ripening, in food processing and as a functional fibre. Like most other polysaccharides, pectins are heterogeneous in terms of chemical structure and molecular weight, and their composition varies according to the source of the botanical tissue and the extraction conditions (ORDONEZ-PEREDA, 2005; THAKUR; SINGH; HANDA, 1997).

The main pectic component is homogalacturan, consisting of alpha-D-galacturonic acid residues linked by alpha-(1,4) glycosidic bonds. The carboxy group of galacturonic acid residues can be esterified with methyl groups, with the degree of esterification changing significantly depending on the degree of differentiation of the cells in the presence of certain pathological conditions such as infection by phytopathogenic fungi. Not only the number, but also the distribution of galacturonic acid carboxy groups along the homogalacturan chain have an influence on the properties of pectin and the firmness of the cell wall (THAKUR; SINGH; HANDA, 1997).

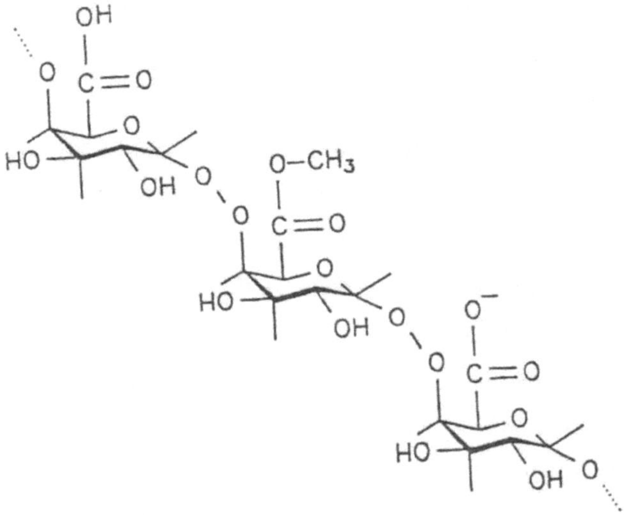

Figure 1 - Chemical structure of a segment of the pectin molecule.
Source: Thakur, Singh and Handa (1997).

2.1.2 Degree of esterification

The degree of esterification corresponds to the number of moles of methanol per 100 moles of galacturonic acid (GUILLOTIN et al., 2005). Determining this percentage requires measurements of the methoxy and anhydrogalacturonic acid content. The degree of esterification and the molar mass of the pectin have an influence on the properties of the gel. The speed of gel formation is related to the degree of esterification, while the strength of the gel depends on the size of the molecule (KLIEMANN, 2006).

Pectins with a high degree of esterification have more than 50% esterified carboxylic groups and form gels with a solids content above 60% and at pHs between 2.8-3.6. Pectins with a low degree of esterification have less than 50% esterified carboxylic groups and have the property of forming thermo-reversible gels by interacting with calcium over a wide range of soluble solids (10-80%) and pH (2.6-6.0). As the final pH must be between 3.0-3.2 in order to achieve adequate gelling and as the natural pH of fruit does not always correspond to this value, it is necessary to adjust it by adding acids or buffering salts permitted by legislation (DZIEZAK, 1991; THAKUR; SINGH; HANDA, 1997).Commercially, highly esterified pectins are described as slow, medium or fast solidifying, referring to the **gelling** time; **"fast solidifying" pectins have a degree of esterification of approximately 75 per cent, while "slow solidifying" pectins have a degree of esterification of 60 per cent. The applications for each type vary** according to the processing. Generally, slow-setting pectins are used in confectionery and fast-setting pectins in more specialised applications, such as the uniform distribution of fruit pieces in jellies (DZIEZAK, 1991).

2.1.3 Extraction methods

Pectic substances are generally extracted by chemical and enzymatic methods. Pectin extraction is a multi-stage physical and chemical process in which the hydrolysis and extraction of macromolecules from plant tissue and their solubilisation are influenced by various factors such as temperature, pH, type of acid and extraction time

(PAGÁN et al., 2001). Commercially, pectin is extracted by treating the raw material with an inorganic acid solution with a pH of around 2.0, under heating. The extract obtained undergoes filtration, clarification and concentration processes and is then commercialised in liquid form. Pectin powder can be produced by mixing the liquid extract with alcohol, precipitating, drying and grinding the pectin (MAY, 1990).

Several studies have found that the type of acid interferes with the pectin extraction process. Canteri-Schemin et al. (2005) used seven types of acid (citric, phosphoric, hydrochloric, nitric, malic, tartaric and sulphuric) to evaluate the extraction of pectin from apple pomace; of the acids studied, citric acid showed the highest extraction yield. The extraction was carried out under reflux at 97 °C for 30 min (solute/solvent 1:50), pH 2.5 for each type of acid, and precipitated with ethanol (95 °GL). Fertonani et al. (2006) also used inorganic and organic acids to extract pectin from apple pomace. The acids used in the extraction were nitric, hydrochloric, succinic, lactic, trichloroacetic, citric and malic. The extraction conditions were boiling for 10 min and cooling in an ice bath at the end. The authors distinguished two types of behaviour of the acidic agents: in inorganic acids, release of pectin at low acid concentrations and destruction at higher levels, and in organic acids, increasing gravimetric yield, indicating aggregation of organic groups in the structure. Nitric acid showed the best gravimetric yield [20 g $(100\ g)^{-1}$] at a concentration of 100 mMol $(L)^{-1}$, in a solute:solvent ratio of 1:40.

Kalapathy and Proctor (2001) investigated the effect of acid concentration on the extraction of pectin from soya husks and the influence of the pH of the precipitation solvent on the yield and purity of the pectin obtained. The authors established the optimum condition for extracting pectin from soya husks as the use of 0.05 N hydrochloric acid and precipitation with alcohol at a pH of 3.5. The researchers observed that concentrated acid solutions can induce an increase in the solubility of smaller pectin molecules, due to partial hydrolysis, and these may not precipitate with the addition of alcohol, resulting in a lower yield. In addition, more acidic precipitation media (pH around 2.0) favour precipitation of components other than pectin, reducing the purity of the product. These studies suggest that both the type of acid and the concentration used to extract the pectin are important factors in the process.

On the other hand, the temperature of the extraction process also influences the yield of pectin. Pagán et al. (2001) extracted pectin from peach residue stored under refrigeration. The extraction was carried out using nitric acid in a jacketed reactor at 60 °C and 80 °C and pH ranging from 1.20-2.53 for 80 min. The pectin was precipitated with 96% ethanol and dried in a vacuum oven until constant weight. The authors found that pH and temperature had a notable influence on pectin extraction yield. For a constant time, with increasing temperature and decreasing pH, the yield of pectin extract increased. They obtained the highest pectin yield at 80 °C for 60 min of extraction at pH values lower than 1.50.

Pectins from the leaves of Cissampelos pareira, a climbing plant from Thailand, were extracted under conditions of pH between 3.0-7.0, temperatures between 25-90 °C, times between 20-60 min and precipitation in 95% ethanol. The optimum conditions for pectin extraction were a solid:solvent ratio of 1:50, in distilled water at 25-28 °C and pH 3.8-4.0, for 60 min. The extracted pectin was of low esterification, consisting of 70-75% uronic acid (SINGTHONG et al. 2005).

Levigne et al. (2002) evaluated parameters such as pH, temperature and type of acid in the extraction of beetroot pectin. The type of acid used, hydrochloric or nitric, did not influence the characteristics of the extracted pectins. The galacturonic acid content remained constant under all extraction conditions, unlike the degrees of esterification and acetylation, which showed variations.

Yapo et al. (2007) studied the effect of extraction conditions with sulphuric acid on the yield and functional properties of beetroot pectin, and observed that the most appropriate conditions for obtaining pectin with the highest degree of purity were pH 1.5 and 80 °C for one hour.

Calliari (2004) used the commercial polygalacturonase enzyme obtained from the fungus Aspergillus niger and the one produced in the laboratory from the yeast Kluyveromyces marxianus to extract pectin from orange pomace. The extraction yield with the laboratory enzyme was higher (15.61 per cent) than with the commercial enzyme (4.54 per cent). The differences were related to the nature of the source of the enzyme. Apparently, the Aspergillus niger enzyme favoured pectin degradation and not extraction, thus resulting in a lower yield.

Other researchers have studied pre-treatment conditions for pectin extraction. Kratchanova; Pavlovaa; Panchev (2004) and Liu; Shi; Langrish (2006) observed that pre-treatment with microwaves increases the porous characteristics of the cell wall capillaries, as well as inhibiting the activity of the enzyme pectinesterase, preventing the degradation of pectin by enzymatic action. These changes can increase the yield of pectin during extraction, as well as favouring gel strength. According to Fishman et al. (2006), lime pectin molecules became less compact with longer microwave heating times, facilitating extraction.

Liu, Shi and Langrish (2006) studied the extraction of pectin from the peel (albedo and flavedo) of oranges. For the flavedo part, the authors used pressure extraction (similar to the technique used for citrus oil extraction), microwave extraction and a combined pressure and microwave process. For the albedo, the Soxhlet and microwave extraction methods were used. The microwave method produced the highest extraction yield. The microwave extractions of flavedo and albedo showed that pectin is found mainly in the albedo, with flavedo accounting for 27% of the total extracted. The combined pressure and microwave method proved to be better at extracting pectin from flavedo than the separate methods. The authors suggest that microwaves help to break down the cell structure, facilitating subsequent pectin extraction.

Zhongdong et al. (2006) established parameters by which pectin can be extracted in a shorter time and with better quality. The work focussed on extracting pectin from orange peel under microwave conditions. Slices of orange peel were immersed in a solution of HCl pH 2 for 10 min. The samples (including the liquid) were heated to 85 °C for 5 min in a microwave oven, at a frequency of 2450 MHZ and power of 1000 W, and filtered. With the use of microwaves, the authors also found that the pectin was extracted more quickly, as the temperature and internal pressure of the cells increased rapidly, favouring the rupture of the cell wall and consequent release of the pectin. The time required for the extraction process was one hour for the classic method and five minutes for the microwave method.

2.1.4 Food sources of pectin

The most common sources for commercial pectin extraction are citrus pulp and apple pomace, citrus fruits being the most widely used (MAY, 1990; THAKUR; SINGH; HANDA, 1997), and for years they have been the main sources of extraction. Currently, there is a great deal of research into the extraction of pectin from different sources, including the utilisation of industrial waste. The pectins obtained have different characteristics, which vary according to their origin, allowing them to be widely used in industry (WILLATS; KNOX; MIKKELSEN, 2006). Table 1 shows the pectin content of different vegetables.

There is a constant search for new sources of pectin, including tomatoes, mango peels and soya beans, among others. According to Sharma, Liptay and Le Marguer (1998), tomato pectin has very promising gelling and emulsifying properties, mainly because it has a high degree of methoxylation (87.8-88.5%). Studies by Berardini et al. (2005) showed that mango peel is also a good source of pectin. Gnanasambandam and Proctor (1999) observed that the use of soya husks, when subjected to a pre-treatment to inactivate enzymes before extraction, favoured an increase in pectin content, obtaining approximately 15% yield.

Table 1. Pectin content of some fruits.

Fruit	% pectin (wet basis)
Apple (Malus spp.)	0,50-1,60
Apple pomace	1,50-2,50
Banana (Musa acuminata)	0,70-1,20
Beetroot pulp (Beta vulgaris)	1,0
Carambola (Averrhoa carambola)	0,66
Carrot (Daucus carota)	0,20-0,50
Lemon (Citrus limon)	2,50-4,0
Mango (Mangifera indica L.)	0,26-0,42
Orange albedo (C. cinesis)	3,50-5,50
Guava (Psidium guajava L.)	0,77-0,99
Papaya (Carica papaya)	0,66-1,0
Passion fruit (Passiflora edullis S.)	0,50
Passion fruit peel	2,10-3,0
Peach (Prunus persica)	0,10-0,90
Pineapple (Ananas comosus L.)	0,04-0,13

Strawberry (Fragaria ananassa)	0,60-0,70
Tamarind (Tamarindus indica L.)	1,71
Tomato *(Lycopersicun esculentum)*	0,20-0,60

Source: Thakur, Singh and Handa (1997).

Singthong et al. (2005) extracted pectin from the leaves of Cissampelos pareira, a climbing plant from Thailand. The pectin extracted was of low esterification, consisting of 70-75% uronic acid. The leaves showed great potential as a source of hydrocolloid, with wide application in the food industry due to their gelling properties.

Guava (Psidium guajava L.) stands out among tropical fruits due to its sensory and nutritional characteristics, being a source of vitamin C, sugars, fibres and minerals (AZZOLINI, 2002). Guava is rich in pectin and its content is influenced by factors such as cultivar, ripeness stage, fruit development period and climatic factors (CERQUEIRA, 2007). The cultivars available in Brazil include: Kumagai, Ogawa, Pedro Sato, Sassaoka, Rica and Paluma. The Pedro Sato cultivar was selected by farmers in the state of Rio de Janeiro and stands out for its large fruit and rough skin (Figure 2). The shape is oblong, the flesh is pink, thick and firm, with a full cavity and few seeds (AZZOLINI, 2002).

Figure 2: Guava cv Pedro Sato.

2.1.5 The importance of pectin in the food industry

For many years the food industry has been looking for technologies to improve food products or to innovate new products in terms of texture, flavour and appearance. Various gelling agents such as carrageenan, alginate, guar and xanthan gums, gelatine, starch and pectins are used to change the texture of industrialised foods. These hydrocolloids are used in different applications since their gelling and thickening properties depend on the product conditions (pH, presence of solutes, salts and temperature). Pectins are mainly extracted from fruit and are therefore natural gelling agents (HAMINIUK, 2007).

As a natural product and due to their different physical properties, pectins are widely used in different foods: jellies, marmalades, yoghurts, desserts, sweets, salad dressings, tomato sauces, among others (PAGÁN; IBARZ, 1999; THAKUR, SINGH; HANDA, 1997). The different types of pectins can be used in different applications.

Pectins with a high degree of esterification are used in sugar-rich products such as jellies (above 60% soluble solids). They can be used in dairy drinks as they prevent casein aggregation during heating at a pH below 4.3 (NAKAMURA et al., 2006).

With the growing demand for low-calorie products in the markets, due to consumers' concern about body weight, jellies with reducing sugars and 30% soluble solids are being produced using low-esterification pectins. These pectins can also be used in bakery products, such as cake fillings, since their gel is thermally reversible, i.e. it will melt and form a good gel again during cooling. In addition, it is less prone to syneresis (MAY, 1990).

The development of fat-free products has increased the use of pectin in products such as salad dressings, ice cream and emulsifying products for meat, replacing fat with pectin (KALAPATHY; PROCTOR, 2001).

Its use in the formulation of dairy products has favoured the development of a wide range of textures in desserts, forming firm or aerated gels due to the calcium present in the milk (MAY, 1990). The stabilisation mechanisms of highly esterified pectins in acidified milk drinks were studied by Nakamura et al. (2006). Pectin added to skimmed milk before acidification absorbed casein micelles on the surface via electrostatic interactions, thus preventing casein aggregation. Laurent and Boulenguer (2003) also used pectin as a stabiliser in dairy drinks, and low-esterification pectin provided the best stabilisation of the drinks. Hansson, Andersson and Leufvén (2001) verified the use of pectin in soft drink stability.

Sharma, Liptay and Le Maguer (1998) showed that tomato pectin has emulsifying properties, facilitating its use in oil/water emulsions and salad dressings. Nachtigall et al. (2004) studied the association of xanthan and carrageenan gums with low-esterification pectins to prevent syneresis in jellies and improve the fragile texture of light jellies, positively influencing the product's acceptability to consumers.

Yapo et al. (2007) evaluated the emulsifying activity and stability of the emulsion prepared with beetroot pectin extracted at pH 1.5, for one hour and at 80 °C. The emulsion was evaluated at temperatures of 4 °C and 23 °C, with 1 and 30 days of storage. The authors concluded that beetroot pectin efficiently favours the stability of oil-in-water emulsions.

Surh, Decker and Mc-Clements (2006) evaluated the influence of pH and the type of pectin (high or low esterification) on the stability and properties of oil-in-water emulsions

stabilised with sodium caseinate. The stability of the emulsions was improved by pectin at pH 5.0, due to the multilayer formed by the pectin. The degree of esterification of the pectin molecules did not influence the stability of the emulsion at the pHs studied.

Mesbahi, Jamalian and Farahnaky (2005) also studied the application of beetroot pectin in food technology. The authors applied beetroot and citrus pectins to ketchup and orange syrup. The use of both pectins increased the viscosity of the ketchup, with 0.7% beetroot pectin increasing the viscosity from 2420 cP to 3923 cP. In orange syrup, 0.7% beetroot pectin increased the viscosity from 135 cP to 253 cP. In ketchup, the pectins had no influence on the colour of the syrups, but the beetroot pectin did influence the final colour of the product. The authors also found that beetroot pectin does not have good gelling properties compared to citrus pectin. However, it can be used as a thickening agent.

Batista, Tanada-Palmu and Grosso (2005) and Giosafatto, Mariniello and Ring (2007) studied the use of low-esterification pectin in biofilms. The use of pectin proved to be viable as it made the film highly soluble.

2.1.6 Functional properties of pectin

A functional food can be defined as one that contains substances with their own biological activity and that can be involved directly or indirectly in the prevention and treatment of diseases (BRASIL, 2008).

Fibres, generally composed of carbohydrates or carbohydrate derivatives of plant origin, are considered to be biologically active substances that act to prevent certain diseases. Most substances characterised as fibres are non-starch polysaccharides. The organic effects of fibres are distinct and depend on their characteristics (TERPSTRA et al., 1998).

The physiological effects of dietary fibre are due to the composition and physical properties of the polysaccharides present. The absorption of glucose by the body depends on the presence of compounds capable of forming viscous solutions, such as pectin. The presence of dietary fibre in the food we eat slows down the absorption of nutrients and helps us feel full. The insoluble fraction of fibre has no effect on lipid metabolism, while the water-soluble portion has hypocholesterolemic properties (PIEDADE; CANNIATTI-BRAZACA, 2003).

Pectin is considered a water-soluble dietary fibre. Several studies have shown that its intake can reduce serum cholesterol and triglyceride levels in rats as well as in humans. Studies carried out on humans also show that citrus pectin, as well as guar gum and Psyllium (Plantago ovato seed) are hypocholesterolaemic agents (FIETZ; SALGADO, 1999).

Various mechanisms have been proposed to explain the action of soluble, gel-forming fibres, such as pectins, in reducing serum triglyceride levels. Fibres, alone or in combination, can act as follows: (a) by altering the digestion and absorption of dietary lipids and/or increasing the faecal excretion of bile acids and neutral sterols, acting as bile acid sequestrants; (b) by increasing the production of short-chain fatty acids in the colon, due to fermentation; and/or (c) by decreasing the percentage of bile acids in the bile, although they increase that of secondary bile acids. Soluble fibres complex with acids in the small intestine and are recovered in the colon, where they are converted by bacteria into secondary bile acids. There is also an increase in hepatic cholesterol synthesis due to the regulation of total body cholesterol homeostasis (FIETZ; SALGADO, 1999; PIEDADE; CANNIATTI-BRAZACA, 2003; TERPSTRA et al., 1998).

Fietz and Salgado (1999) evaluated the effect of high and low esterification pectin and cellulose on cholesterol and triglyceride levels in rats. The treatments with high-esterification pectin were more effective.

Piedade and Canniatti-Brazaca (2003) studied the effect of pineapple residue and high-esterification citrus pectin on cholesterol levels in rats. Pectin was more effective in reducing total cholesterol and LDL-cholesterol. The reduction in serum levels of the LDL-cholesterol fraction was attributed to the increased intake of soluble fibres, favouring fermentation, which reduces the synthesis of this fraction.

CHAPTER 3

MATERIAL AND METHODS

3.1 MATERIAL

The raw material investigated in this study was guava (Psidium guajava L.) of the red Pedro Sato cultivar. The guavas were purchased in February 2007 from the Gleba orchard in the district of Santa Terezinha, municipality of Itaporã, state of Mato Grosso do Sul, Brazil.

3.2 METHODS

3.2.1 Raw material preparation

Twenty kilos of guavas were selected according to their weight (130 to 140g), size, appearance (healthy fruit), shape (uniform) and degree of ripeness defined by the yellow colour of the skin observed visually in order to obtain more homogeneous samples.

The fruit was washed, sanitised with Sumaveg brand dichloro s. triazinatrione sodium dihydrate at a concentration of 0.66% for 10 minutes and lightly dried with absorbent paper. The fruit was then divided into four batches and manually fractionated with stainless steel knives for subsequent drying: 1) pulp with peel, without core; 2) peeled guavas without core (pulp); 3) peel and 4) core. Five fruits were taken at random from each batch and kept under refrigeration for 24 hours for subsequent physical and chemical characterisation of the fruit in natura.

The fractions from each batch were dried separately in an oven with forced air circulation at 55 °C for 72 hours and ground in a Marconi knife mill until the flours were obtained, which were stored in hermetically sealed glass containers at room temperature.

3.2.2 Physical and chemical characterisation of whole guavas and their fractions *in natura* and of the flours from the fractions obtained

For each batch, two groups of approximately 350g (around 2.5 fruits) of whole guavas were sampled and separated into peel, pulp with peel and fresh core. They were then analysed for pH, titratable acidity and soluble solids content, while the fresh fruit pulp samples, in addition to the aforementioned analyses, were assessed for total and reducing sugar content.

The pHs of the samples were measured using direct readings in a digital potentiometer, Mettler Toledo model pH320 (Brazil), previously calibrated with standard solutions of pH 4.0 and 7.0, using the whole fruit or its fractions that had been liquefied and homogenised at a controlled temperature of 25 °C. The soluble solids content was measured by direct reading on a digital refractometer, Carlzeissjena brand, model I, precision 0.1 °Brix, while the titratable acidity values were determined by titration with a 0.1 N NaOH solution (INSTITUTO ADOLFO LUTZ, 2005). Total and reducing sugar levels were determined using the Lane Eynon chemical method, based on copper oxidation (RANGANNA, 1977).

The different flours were analysed as follows. Moisture content was determined by drying the samples in an oven at 105 °C until constant weight, ash was quantified using the incineration method at 550 °C and titratable acidity by titration with a 0.1 N NaOH solution. All the analyses were carried out according to the techniques described by the Adolfo Lutz Institute (2005). The total sugar and reducing sugar contents were determined using the Lane Eynon chemical method based on copper oxidation (RANGANNA, 1977). Water activities were determined using direct measurements in an Aqualab hygrometer, model CX-2T (Aqualab - Decagon Devices Inc., USA) at a temperature of 25 °C. The pHs of the samples were measured by direct reading on a digital potentiometer, Mettler Toledo model pH320 (Brazil), previously calibrated with standard solutions of pHs 4 and 7, after solubilising the samples in distilled water.

The quantification of pectin content followed the methodology described by Ranganna (1977). 25 g of each sample was weighed into a 1000 mL beaker. Extraction was carried out with 400 mL of 0.05 N hydrochloric acid and boiled for one hour, replacing the water lost during evaporation. These materials were cooled and transferred to 500 mL flasks, where the volumes were topped up with water, manually shaken by inverting the flasks for approximately 2 min and filtered through Whatman No. 4 paper.

Two 100 mL aliquots of each filtrate were transferred to two 1000 mL beakers and 300 mL of distilled water and 10 mL of 1 N sodium hydroxide were added to each, stirring continuously, and then left to stand overnight. After this period, 50 mL of acetic acid solution was added, and five minutes later, 50 mL of calcium chloride solution was added while stirring.

After an hour's rest, the materials were brought to the boil for 1 min, then filtered through pre-prepared filter paper (the paper was moistened with hot water, placed in an oven at 105 °C for two hours, cooled in a desiccator and weighed in a lidded container). The precipitate was washed with almost boiling water until all the chloride was removed. The presence of chlorides was checked using a 1% silver nitrate solution. The filter papers containing the precipitates were transferred to lidded containers, which were kept in an oven at 100 °C overnight, cooled in a desiccator and weighed. The amount of calcium pectate was calculated using equation 1.

$$(\%)Peclalcium = \frac{P*500*100}{V*A} \quad \text{Equation (1)}$$

Where: P = mass of calcium pectate (g)

V = volume of filtrate (mL)

A = sample mass (g)

All the physical and chemical analyses were carried out in triplicate.

3.2.3 Experimental planning to optimise pectin extraction yields from guava pulp and guava peel flours

The pectin was extracted in an acidic medium using citric acid. The procedure to optimise the yield of guava pectin extraction was carried out only on the fractions with the highest pectin content. Samples of each flour (4 g) were dissolved in 200 mL of distilled water. The extractions were carried out at a constant temperature (97 °C) and under different experimental conditions of citric acid concentration (3.5-7.0 g. 100 g^{-1}) and time (25-95 min). After acid extraction, the samples were cooled at 4 °C for two hours, filtered through polyester cloth and the contents of the filter discarded. Ethyl alcohol (95%) was added to the filtrate containing pectin in a 1:2 ratio (one part pectin solution and two parts alcohol). After an hour's rest, the pectin precipitated and was separated by filtering through polyester cloth. The pectin obtained was dried in a drying oven at 55 °C until constant weight. The pectin yield was calculated from the initial quantity of raw material used (Equation 2).

To optimise the yields of pectin extractions from guava pulp and guava peel flours, a central composite rotational design was used with four axial points and three repetitions at the central point, totalling 11 trials (RODRIGUES; IEMMA, 2005). The independent variables chosen were citric acid concentration (g L^{-1}) and extraction time (min). The actual original levels of the independent variables were defined after carrying out preliminary tests and are shown together with the coded levels in Table 2. The mass yield of pectin extracted in each experiment was considered the response or dependent variable. The trials were conducted with two repetitions and were randomised.

$$Re\ n\ dim\ ent = \frac{pectin\ extract * 100}{flour\ dough\ (dry)} \qquad \text{Equation (2)}$$

Where: Pectin extracted = mass of pectin obtained after extraction and drying in grams; Mass of flour (dry) = mass of sample on a dry basis in grams.

Table 2. Experimental design matrix.

Essay	Coded variables	Variabl original

	es			
	Citric acid (%)	Time (min)	Citric acid (%)	Time (min)
1	-1	-1	3,5	35
2	1	-1	6,5	35
3	-1	1	3,5	85
4	1	1	6,5	85
5	-1,414	0	3	60
6	1,414	0	7	60
7	0	-1,414	5	25
8	0	1,414	5	95
9	0	0	5	60
10	0	0	5	60
11	0	0	5	60

3.2.4 Titrimetric characterisation of the pectins in guava pulp and guava peel flours

The pectin samples obtained under the optimum experimental conditions were quantified by titration (WANG; PAGAN; SHI, 2002 apud FERTONANI, 2006). Approximately 250 mg of pectin was moistened with 2 mL of ethyl alcohol P.A. and solubilised in 25 mL of deionised water under constant stirring for 30 min on a magnetic stirrer, after which the pH of the solution was determined. The free carboxyls of the anhydrogalacturonic acids were neutralised with a 0.1 N NaOH solution. The esterified carboxyls after saponification with 10 mL of 0.25 N NaOH solution for 30 min at room temperature were neutralised with 10 mL of 0.25 N HCl solution and again neutralised with 0.1 N NaOH solution, thus obtaining the values of mEq of NaOH for the two types of carboxyls, free and esterified, respectively represented **by mEq' and mEq". With the data obtained, calculations were carried out to determine the** amount of mass per mEq (Equation 3), galacturonic residue content (Equation 4), methoxyl content (Equation 5), percentage of polygalacturonic acids (Equation 6), percentage of neutral sugars (Equation 7) and degree of esterification (Equation 8).

Amount of mass per mEquivalent

$$Z = \frac{massapectin}{mEqtotals} \qquad \text{Equation (3)}$$

Galacturonic residue content

$$AUA\% = \frac{17600}{Z} \qquad \text{Equation (4)}$$

Methoxyl content

$$MeO\,\% = \frac{(mEq'' * 31 * 100)}{massapectin} \qquad \text{Equation (5)}$$

Percentage of polygalacturonic acid	*Acid fraction* = *AUA%* + *MeO* %	Equation (6)
Percentage of neutral sugars	*Other fraction* = 100 - *acid fraction*	Equation (7)
Degree of esterification	$DE = (\frac{176}{31}) * (\frac{MeO}{AUA})$	Equation (8)

3.2.5 Analysing the results

The data obtained on the physical and chemical characterisation of the raw material, its fractions and flours were used to calculate the mean, standard deviation and coefficient of variation. Analyses of variance and the Tukey test at a 5% probability of error were also carried out to compare the means obtained.

The pectin yield data obtained was subjected to analysis of variance and multiple regression analysis to draw up mathematical models with the coded variables, using the SAS application (THE SAS SYSTEM, 1992).

The mathematical models for analysing the results follow equation (9):

$$y = b_0 + b_1x_1 + b_2x_2 + b_{11}x_1^2 + b_{22}x_2^2 + b_{12}x_1x_2 \quad \text{Equação (9)}$$

Where: y = pectin yield in dry weight;

bi = regression coefficients for linear effects;

bik = interaction regression coefficients;

bii = coefficients for regression of quadratic effects;

xi = experimental coded levels of the variables.

To visualise the effect of the variables on the responses, response surface graphs and Pareto diagrams were constructed using the Statistica programme (STATISTICA FOR WINDOWS, 1995).

CHAPTER 4

RESULTS AND DISCUSSION

4.1 PHYSICAL AND CHEMICAL CHARACTERISATION OF FRUITS, FRACTIONS AND FLOURS

The average results of the physical and chemical analyses of the whole fruit, peel, pulp, pulp with peel and core of guava, as well as the standard deviations, are shown in Table 3.

Table 3 Means and standard deviations of the physical and chemical characteristics of the whole fruit, skin, pulp, pulp with skin and core of guava[1].

Features	Fruit whole	Bark	Pulp	Pulp w/ bark	Core	CV*
Reducing sugars g (100 g)$^{-1}$	nd	nd	5,90 ± 0,54	nd	nd	-
Total sugars g (100 g)$^{-1}$	nd	nd	8,00 ± 0,49	nd	nd	-
pH	3,78 ^{A} ± 0,02	3,22 ^{D} ± 0,02	3,48 ^{C} ± 0,03	3,49 ^{C} ± 0,02	3,69 ^{B} ± 0,02	0,01
Titratable acidity g (100 g citric acid)$^{-1}$	0,38 ^{B} ± 0,01	0,41 ^{B} ± 0,01	0,50 ^{A} ± 0,03	0,41 ^{B} ± 0,02	0,24 ^{C} ± 0,01	0,04
Soluble Solids (°Brix)	9,20 ^{B} ± 0,20	8,07 ^{C} ± 0,12	8,07 ^{C} ± 0,12	8,07 ^{C} ± 0,12	10,00 ^{A} ± 0,20	0,02

* CV - coefficient of variation nd - not determined

[1] - Means followed by equal superscript letters on the same line do not differ at the 5% probability level using the Tukey test.

The average total sugar content of 8.00 g (100 g)$^{-1}$ and reducing sugar content of 5.90 g (100 g)$^{(-1)}$ for guava pulp obtained in this study were higher than those found by Vila et al. (2007), respectively 6.15 g (100 g)$^{-1}$ and 5.16 g (100 g)$^{(-1)}$) for the Pedro Sato cultivar. Mattiuz and Durigan (2001) obtained 2.09 g (100 g)$^{-1}$ of reducing sugars for the same cultivar; these differences can be attributed to different degrees of fruit ripeness or climatic conditions. Sato, Sanjinez-Argandona and Cunha (2004) studied Paluma guavas and quantified 10.52 g (100 g)$^{-1}$ of total sugars, suggesting that in addition to the difference in degree of ripeness and soil and climate conditions, the cultivar also influences the fruit's centesimal composition. According to Bashir, Abu-Bakr and Abud-Goukh (2003 apud SANJINEZ-ARGANDONA, 2005), the high sugar content can be attributed to the increase in enzymatic activity responsible for hydrolysis of starch and also to the decline in the speed of sugar breakdown during fruit respiration.

The average pH of the fresh fruit samples and the fractions studied ranged from 3.28 to 3.78 and differed ($P < 0.05$) from one another, with the exception of the pulp and pulp with peel fractions, which did not differ ($P > 0.05$). The highest average pH value was observed in the whole fruit, while the peel had the lowest value. Brunine, Oliveira and Varanda (2003) found a pH of 3.86 for the pulp of guava cv Paluma; Sato, Sanjinez-Argandona and Cunha (2004) also obtained pH values in the 3.6 to 4.0 range for the pulp of the same cultivar. The difference observed may be associated with differences between the cultivars. Hojo et al. (2007) found average pH values of 3.90 to 4.28 for whole fruit from Pedro Sato guavas at different pruning times. It can therefore be inferred that the time of pruning, degree of ripeness, soil type and rainfall during the harvest period can influence the physical and chemical characteristics of the fruit.

The average titratable acidity values obtained in this study ranged from 0.24-0.50, being lower for the core and higher for the guava pulp. The titratable acidity of the peel, pulp with peel and whole fruit were similar ($P > 0.05$) to each other, but differed ($P < 0.05$) from the core and pulp. The average values obtained for soluble solids and titratable acidity for the core were 10.00 °Brix and 0.24 g $(100\ g)^{-1}$, respectively, which suggests that this fraction is sweeter to the taste and less acidic than the other fractions. In relation to the pulp, the higher acidity observed can be attributed to the greater presence of ascorbic acid, in addition to citric acid (SANJINEZ-ARGANDONA, 2005). In whole fruit, the value obtained of 0.38 g $(100\ g)^{-1}$ was lower than the 0.54 g $(100\ g)^{-1}$ found by Azzolini, Jacomino and Spoto (2004) for the same cultivar. However, Mattiuz and Durigan (2001) determined a value close to titratable acidity for pulp with peel of 0.44 g $(100\ g)^{-1}$ in Pedro Sato guavas. Brunine, Oliveira and Varanda (2003) found an average titratable acidity of 0.41 g $(100\ g)^{-1}$ for Paluma guava pulp, while Jacomino et al. (2003) found 0.62 g $(100\ g)^{-1}$ for Pedro Sato guava pulp. The variation observed between the results obtained in this study and those in the literature can be attributed to the degree of ripeness, as well as the difference between the cultivars and soil and climate conditions.

The soluble solids obtained ranged from 8.07-10.00 °Brix and did not differ ($P > 0.05$) between the skin, pulp and pulp with guava skin, but they did differ ($P < 0.05$) from the values observed for the whole fruit and the core. This variation was to be expected given that the core is generally more appreciated than the pulp precisely because it is sweeter. In relation to the whole fruit, Jacomino et al. (2003) reported a value close to soluble solids (9.38 °Brix) for the same cultivar. Mattiuz and Durigan (2001) determined a similar average soluble solids value (8.03 °Brix) for Pedro Sato guava pulp with peel, while Azzolini, Jacomino and Spoto (2004) reported lower average soluble solids values (7.84 °Brix) for Pedro Sato guava pulp.

The appearance of the flours obtained from guava peel, pulp with peel, pulp and core can be seen in Figure 3 and the physical and chemical characteristics in Table 4. Few studies have been carried out on flours obtained from fruit fractionation and none have worked with guava, which encouraged the study of flours from the peel, pulp with peel, pulp and core of guava.

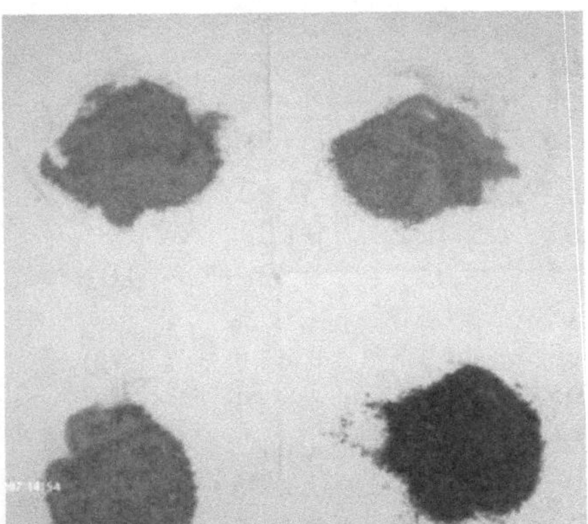

Figure 3: Flours of guava pulp with peel, guava pulp, guava core and guava peel (clockwise from top left)

The average moisture contents determined for the different flours ranged from 10.52-13.24 g (100 g)$^{-1}$, and these did not differ ($P > 0.05$) from each other, except for guava peel flour, which had the lowest average. The moisture content of the guava flours analysed is within the limits set by the National Health Surveillance Agency (ANVISA), which establishes a maximum of 15% (m/m) moisture for flours (BRASIL, 2007). As the moisture content of the flours studied was less than 13.50 g (100 g)$^{-1}$, it can be considered that these products have good physical and chemical stability, as long as they are stored properly in hermetically sealed packages (FERTONANI, 2006).The average water activity fluctuated in a range of values between 0.37-0.45, differing ($P < 0.05$) between the flours evaluated. Foods with a water activity of less than 0.60 are microbiologically stable (SANJINEZ-ARGANDONA, 2005). The water activity value of guava peel flour of 0.37 was the lowest among the flours analysed, due to the lower amount of water present in the peel compared to the pulp and core (Table 4).

Table 4 Means and standard deviations of the physical and chemical characteristics of guava peel, pulp, pulp with peel and core flours.

Features	Bark Flour	Pulp flour	Pulp Flour with Peel	Crumb flour	CV
Humidity[1] [g (100 g)$^{-1}$]	10,52 ^{B} ± 0,75	12,55 ^{A} ± 0,44	13,24 ^{A} ± 0,28	12,51^{A} ±0,62	0,04
Water activity	0,37 ^{D} ± 0,01	0,40 ^{B} ± 0,01	0,45 ^{A} ± 0,00	0,39 ^{C} ± 0,00	0,01
Ashes [g (100 g)$^{-1}$]	3,38 ^{A} ± 0,26	1,89 ^{B} ± 0,38	2,69 ^{A} ± 0,24	3,59 ^{A} ± 0,37	0,11
Total sugars [g (100 g)$^{-1}$]	30,64 ^{A} ±0,91	31,45 ^{A} ± 1,05	33,12 ^{A} ± 1,72	31,90 ^{A} ±1,22	0,04
Reducing sugars [g (100 g)$^{-1}$]	9,57 ^{A} ± 0,35	9,83 ^{A} ± 0,15	9,03 ^{A} ± 0,10	9,32 ^{A} ± 0,24	0,02
pH	4,20 ^{C} ± 0,02	4,28 ^{B} ± 0,02	4,29 ^{B} ± 0,01	4,45 ^{A} ± 0,01	0,00
Total Acidity [g (100g citric acid)$^{-1}$]	2,04 ^{B} ± 0,00	2,27 ^{A} ± 0,04	2,12 ^{B} ± 0,05	1,64 ^{C} ± 0,06	0,02
Pectin [g calcium pectate (100g)$^{-1}$]	0.85 ^{A,B} ± 0,06	1,02 ^{A} ± 0,12	1,02 ^{A} ± 0,11	0,68 ^{B} ± 0,07	0,10

CV - coefficient of variation

[1] - Averages followed by equal superscript letters in the line do not differ at the 5% probability level using the Tukey

test.

The average total mineral content of the flours, determined by ash analysis, was in the range 1.89-3.59 g (100 g)$^{-1}$ and did not differ significantly ($P > 0.05$) from each other, with the exception of guava pulp flour, which was lower ($P < 0.05$) than the others. The ash content of guava peel flour was similar to that found by Kliemann (2006) for yellow passion fruit peel flour with flavedo of 3.36 g (100 g)$^{-1}$. This author compared the results of yellow passion fruit flour with other commercial flours, concluding that the product is a good source of minerals. Studies carried out with wheat flour (GUTKOSKI; NODARI; JACOBSEN-NETO, 2003) and cassava flour (CHISTÉ et al. 2007) showed ash contents of 0.50 g (100 g)$^{-1}$ and 0.54 g (100 g)$^{(-1)}$), respectively. In a study of green banana flour, Fasolin et al. (2007) reported ash values of 2.62 g (100 g)$^{-1}$, close to the average content found for guava pulp and peel flour. From the above, considering the results achieved in this study, it can be concluded that they can all be considered good sources of minerals.

The average content of reducing sugars ranged from 9.03-9.83 g (100 g)$^{-1}$, while total sugars ranged from 30.64-33.12 g (100 g)$^{-1}$, which means that the flours obtained from different guava fractions have considerable energy value. The levels of reducing and total sugars did not differ ($P > 0.05$) between the flours studied. The total sugar content determined for the guava flours was lower than the 33.44 g (100 g)$^{-1}$ reported by Fertonani (2006) for Joaquina apple pomace flour, and 32.54 g (100 g)$^{-1}$ verified by Canteri-Schemin (2003) for Belgolden apple, except for the flour from guava pulp and peel, which obtained a higher value. Only traces of total and reducing sugars were found in green banana flour (FASOLIN et al., 2007). Considering that total sugars sensorially represent the sweetness of the fruit, it is expected that apples and guavas will have a higher sugar content than green bananas, due to their stage of ripeness. With regard to the content of reducing sugars, Canteri-Schemin (2003) found a value of 30.20 g (100 g)$^{(-1)}$) in apple pomace flour, much higher than the values observed in the guava flour described in this study.

Total sugars are made up of reducing sugars (monosaccharides) and non-reducing sugars (polysaccharides) that have been previously hydrolysed (RIBEIRO; SERAVALLI, 2004). Therefore, according to the data obtained, it can be assumed that approximately two thirds of the sugars found in the guava flours studied are non-reducing sugars, probably sucrose.

The flours had a pH ranging from 4.20 to 4.45, which are considered acidic products (pH 4.0-4.5), which allows for greater stability in relation to deterioration by pathogenic microorganisms (JAY, 2005). The pH differed ($P < 0.05$) between the flours, with the exception of the pulp and pulp-with-peel flours, which did not differ ($P > 0.05$). The lowest pH was observed in the peel flour and the highest in the guava core flour. Miranda and El-Dash (2002) found an average pH value of 6.2 for whole wheat flour, which was higher than that found for guava fraction flour. It is to be expected that the guava values would be higher, since guava has a lower pH than wheat.

The titratable acidity ranged from 1.64-2.27 g (100g citric acid)$^{-1}$. The lowest average was observed in the crumb flour and the highest in the guava pulp flour, which was to be expected given the average pH values obtained for these samples. The titratable acidity averages differed ($P < 0.05$) between the flours, with the exception of the peel and pulp with peel flour averages, which did not differ ($P > 0.05$) between them. Fertonani (2006) reported an acidity of 1.0 g (100g citric acid)$^{-1}$ for apple pomace flour cv. Joaquina, a value lower than those found for guava fraction flours. CHISTÉ et al. (2006) analysed ten samples of cassava flour from the seca group and obtained total acidity values in the range of 4.11 to 7.10 meq NaOH (100g)$^{-1}$. CHISTÉ et al. (2007) found an average value of 10.19 meq NaOH (100g$^{)(-1)}$ for cassava flour

The average pectin content of the flours analysed was in the range 0.68-1.02 g (100 g)$^{-1}$. The highest pectin contents were observed in the pulp and pulp with peel flours, which were equal but differed ($P < 0.05$) from the guava core flour, which had the lowest content. The pectin content of the peel flour was intermediate to the levels found for the pulp, pulp with peel and core flours, and did not differ significantly ($P > 0.05$) from them. Linhares et al. (2007), studying Pedro Sato guava pulp, reported a dry basis total pectin content of 0.88 g (100 g)$^{-1}$, Vila et al. (2007) values of around 0.66 g (100 g)$^{-1}$ and Xisto et al. (2004) pectin values of 0.4 g (100 g)$^{-1}$ for the same cultivar. Considering that guava pulp flour has an average moisture content of 12.55 g (100 g)$^{-1}$, the average pectin value obtained is close to that found by Linhares et al. (2007), and higher than that obtained by Vila et al. (2007) and Xisto et al. (2004).

4.2 PECTIN EXTRACTION

Figures 4 and 5 illustrate the process of extracting pectin with citric acid and its alcoholic precipitation, respectively. The extraction of pectin from different raw materials using citric acid has been frequently used in various studies (CALLIARI, 2004; CANTERI-SCHEMIN, 2003; KLIEMANN, 2006; MARCON et al., 2005). Figure 6 (a) shows pectin in gel form, obtained after extraction and precipitation, and Figure 6 (b) shows pectin dried at 55 °C.

Figure 4 - Acid extraction of pectin.

Figure 5: Alcoholic precipitation of pectin.

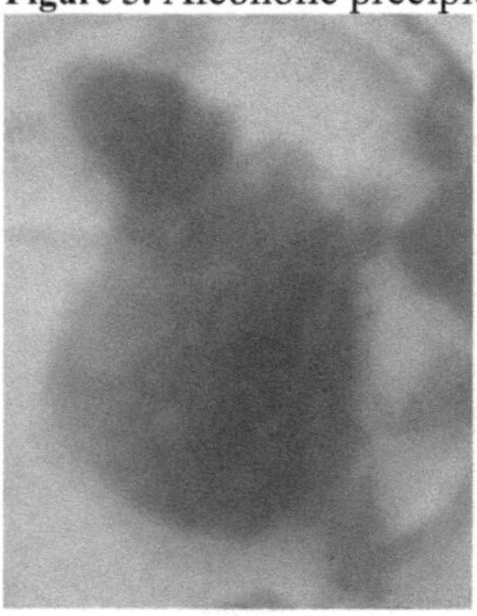

(a)

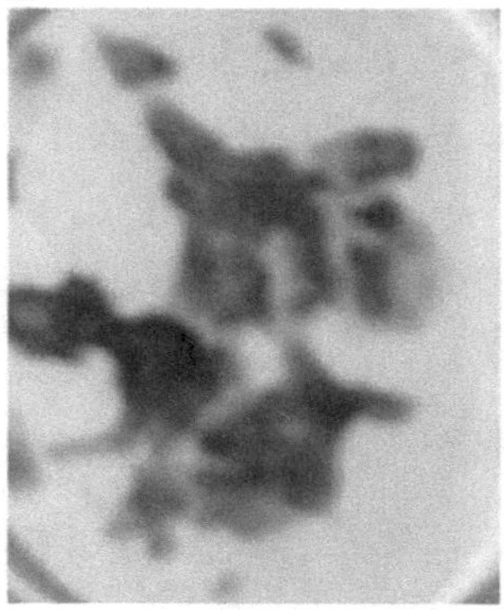

(b)

Figura 6. Pectin extracted wet (a) and dry (b).

6.2.1 Optimising the extraction of pectin from guava pulp

Table 5 shows the average yields of pectin extracted from guava pulp flour.

Table 5. Experimental planning and average yield of pectin extracted from guava pulp flour *cv* Pedro Sato

Experiment	Coded variables		Original variables		Pectin yield [g (100 g)$^{-1}$]		
	Acid	Time	Acid [g (100 g)$^{-1}$]	Time (min)	Repetition 1	Replay 2	Average[1]
1	-1	-1	3,5	35	7,64	8,02	7,83 C
2	1	-1	6,5	35	13,37	13,91	13,64 A
3	-1	1	3,5	85	9,28	9,47	9,38 B
4	1	1	6,5	85	13,09	12,70	12,90 A
5	-1,414	0	3	60	9,82	9,31	9,56 B
6	1,414	0	7	60	13,85	13,48	13,66 A

7	0	-1,414	5	25	8,23	7,59	7,91 [C]
8	0	1,414	5	95	12,93	12,62	12,77 [A]
9	0	0	5	60	12,76	12,28	12,52 [A]
10	0	0	5	60	12,57	12,93	12,75 [A]
11	0	0	5	60	12,73	12,15	12,44 A

[1]- Means followed by equal superscript letters in the column do not differ at the 5% probability level using the Tukey test.

The average yields of pectin extracted from guava pulp flour cv Pedro Sato ranged from 7.83-13.66 g $(100\ g)^{-1}$. The yields obtained in experiments 2, 4, 6, 8, 9, 10 and 11 (citric acid concentrations of 5 to 7 g $(100\ g)^{-1}$) did not differ ($P > 0.05$) from each other, but did differ ($P < 0.05$) from the others. The highest pectin yield values were observed in these experiments. Experiments 3 and 5 also did not differ ($P > 0.05$) from each other, as did experiments 1 and 7, which had the lowest yields.

The analysis of variance and the effects of citric acid concentration and time on the yield of pectin extraction from guava pulp flour are shown in Table 6 and Figure 7.

Table 6. Analysis of variance of pectin extraction yield from guava pulp flour cv Pedro Sato as a function of citric acid concentration [g (100 g)-1] and extraction time (min).

Factor analysed	Sum of squares	Degrees of freedom	Mean square	F	P
(1) Acid (L)	**57,98**	**1**	**57,98**	**71,60**	**0,0001**
Acid (Q)	2,64	1	2,64	3,26	0,0896
(2) Time (L)	**14,63**	**1**	**14,63**	**18,07**	**0,0006**
Time (Q)	**14,15**	**1**	**14,15**	**17,98**	**0,0006**
1L by 2L	2,62	1	2,62	3,24	0,0907
Lack of adjustment	**10,27**	**3**	**3,42**	**1121,22**	**0,0001**
Waste	10,78	13	0,674		
Total	102,80	21			

Coefficients in bold are significant.

The analysis of variance showed that the linear effects of the variables: acid concentration and extraction time, and the quadratic effect of time on pectin extraction yield were significant, so both acid and extraction time influence pectin extraction yield. To better visualise the effects of the variables and interactions, as shown in Table 6, the graph in Figure 7 was constructed.

All the effects with values greater than 2.01 ($P = 0.05$), which are located to the right of the dashed line, were significant. Thus, it can be seen that the factors: acid (L) and time (L) have a positive effect on pectin extraction yield, i.e. an increase in either of these factors leads to an increase in yield, while the effect of the time factor (Q) has a negative influence. Conversely, an increase in this factor leads to a decrease in yield. These statistical results are consistent, as long extraction times favour the degradation of the pectin molecule, especially when associated with a high acid concentration (KALAPATHY; PROCTOR, 2001). Canteri- Schemin (2003) found similar behaviour of the independent variables (citric acid concentration and extraction time) for the extraction of apple pomace pectin with citric acid, in which the linear effects were positive and the quadratic effects negative. Calliari (2004) obtained a positive linear effect for the acid and a positive quadratic effect for the extraction temperature when extracting orange pomace pectin with citric acid, while only the temperature had a significant effect when extracting with acetic acid.

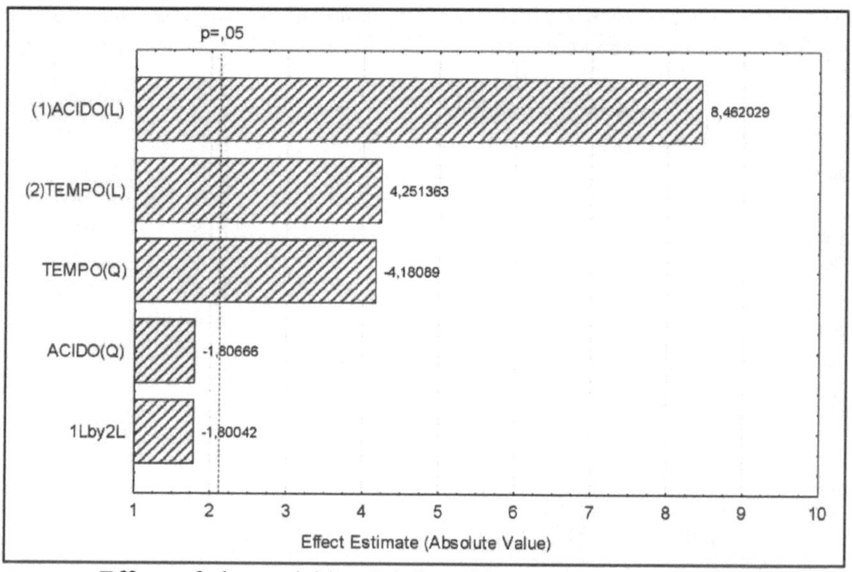

Figura 7. Effect of the variables "citric acid concentration" and "time" and their interactions on the yield of pectin extraction from guava pulp flour cv Pedro Sato.

The quality of the model obtained with citric acid can be seen in Figure 8, which shows the relationship between the standardised effects and the expected values.

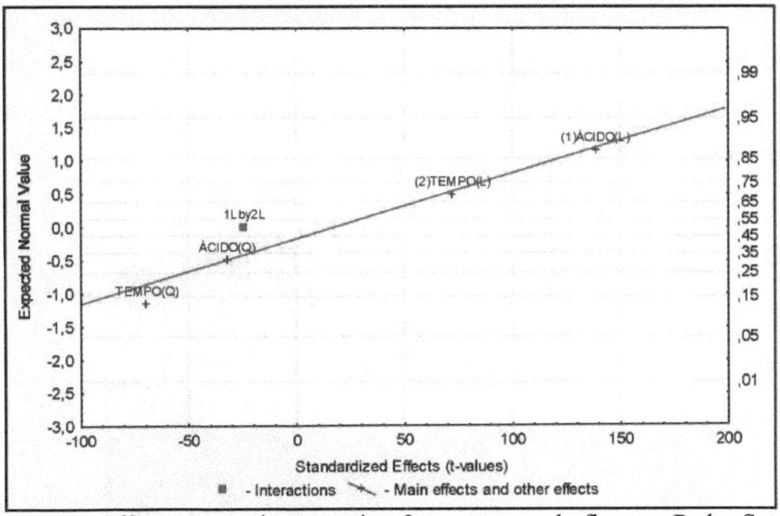

Figura 8. Effects on pectin extraction from guava pulp flour cv Pedro Sato

Table 7 shows the analysis of variance of the adjusted model.

Table 7. Analysis of variance for the adjusted model of pectin extraction yield from guava pulp flour cv Pedro Sato as a function of citric acid concentration [g (100 g)$^{-1}$] and extraction time (min).

Factor analysed	Degrees of freedom	Sum of squares	Mean square	F	P
Model	3	83,8625	27,9542	26,63	0,0001
Error	18	18,8986	1,0492		
Corrected total	21	102,7611			

The adjusted model was significant (P=0.0001), and the coefficients of determination (R^2) and variation (CV) were 0.8161 and 0.0899 respectively, showing that 81.6% of the variation in the response is explained by the model. The F value of the adjusted model was at least five times higher than the tabulated F value, which confirms that the model can be predictive for the levels established in this planning (RODRIGUES; IEMMA, 2005). Equation 10 shows the adjusted model built for pectin yield from guava pulp flour. The model was adjusted by eliminating the non-significant factors: acid (Q) and interaction of acid (L) vs time (L).

$$Rend = 12{,}11 + 1{,}89 * Ac + 0{,}96 * t - 0{,}98 * t^2 \quad \text{Equação (10)}$$

Where:

Rend = pectin extraction yield [g (100 g)$^{-1}$];

Ac = citric acid concentration [g (100 g)$^{-1}$];

t = extraction time (min).

Figures 9 and 10 show the 2D and 3D response surfaces for obtaining the optimum process condition based on the regression model of the pectin extraction yield from guava pulp flour as a function of citric acid concentration and time, keeping the extraction temperature fixed (97 °C).

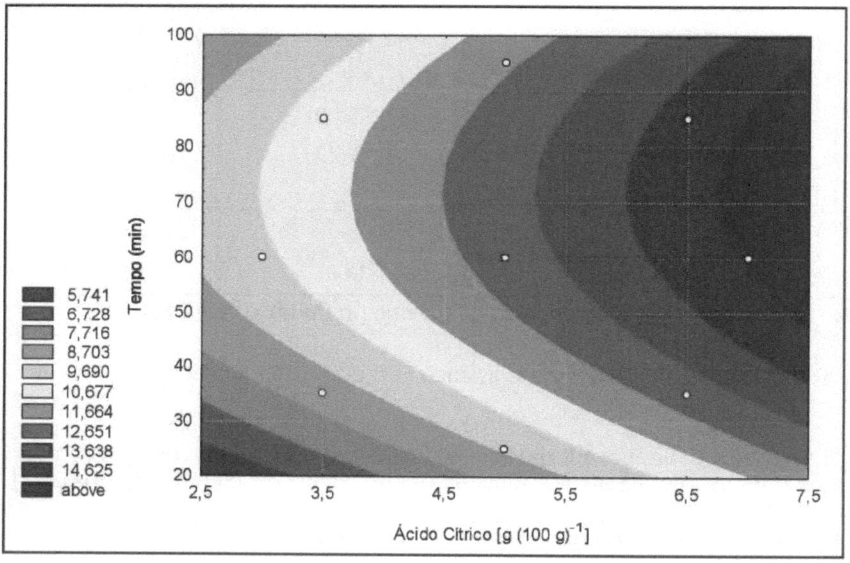

Figura 9. Contour plot of the yield [g (100 g)$^{-1}$] of pectin extraction from guava pulp flour cv Pedro Sato as a function of time (min) and citric acid concentration [g (100 g)$^{-1}$].

Figure 9 shows that the yield increases with increasing citric acid concentration and extraction time of up to 70 min. This can also be seen in Figure 10, which in addition to proving this trend, shows that by setting the time in the 60 min range and increasing the acid concentration to its highest value, the yield increases, while setting the concentration in the lower range and increasing or decreasing the extraction time, the yield decreases, indicating that the effect of the acid concentration is critical in this process. Thus, it can be seen in Figure 10 that there is a plateau, and not just one point, where the extraction yield is optimal, showing the region covered by the experiment.

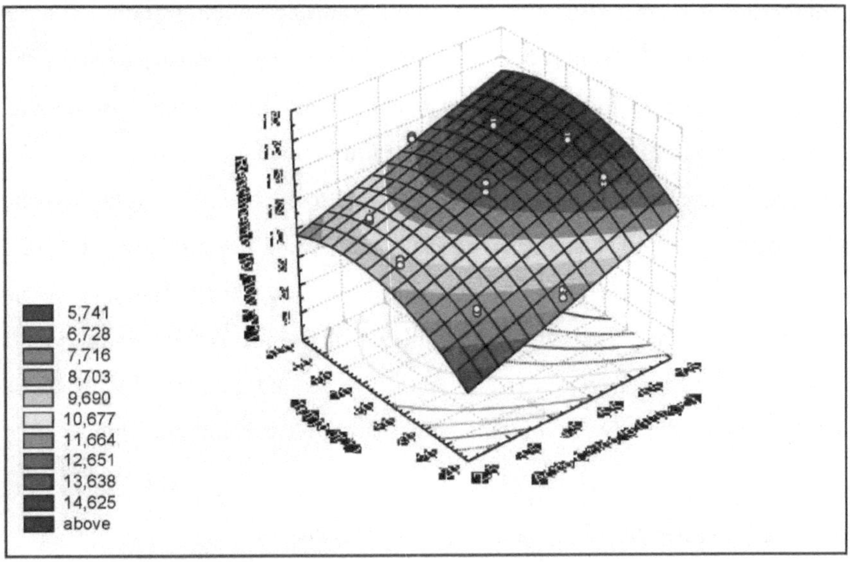

Figure 10. Response surface plot of the yield [g (100 g)$^{-1}$] of pectin extraction (legend) from guava pulp flour cv Pedro Sato as a function of time (min) and citric acid concentration [g (100 g)$^{-1}$].

The citric acid concentrations in experiments 2, 4, 6, 8, 9, 10 and 11 promoted higher pectin extraction yields, which were statistically similar to each other, with only the extraction time varying. However, as discussed above, relatively long extraction times can degrade pectin by reducing the amount to be precipitated with alcohol due to the formation of a weak gel, which causes a reduction in the final pectin yield (PAGÁN et al., 2001), just as higher concentrations of acids can degrade pectin. In the case of citric acid, according to Fertonani et al. (2006) pectin extracted with high concentrations of this acid may also contain other compounds or the citric acid may bind to the pectic structure precipitated in ethanol. Canteri-Schemin et al. (2005) in the extraction of pectin from apple pomace using tartaric, malic, citric, phosphoric, sulphuric, hydrochloric and nitric acids obtained better yields from citric acid at 97 °C in a time interval ranging from 10 to 210 min and concentration from 0.05 to 9.95 g (100 mL)$^{-1}$, the theoretical optimum condition established by the model was 6.2 g (100 mL$^{)(-1)}$ and 150 min.

In this study, the theoretical condition was established using the model generated through the response surface (Equation 10), with the citric acid concentration of 6.7 g (100 g)-1 and time of 70 min being considered the best combination within the maximum extraction range. In the tests carried out based on the planning, it was found that the best gel formation was obtained at a concentration of 5 g (100 g)$^{-1}$ of citric acid and an extraction time of 60 min, as visually observed by its firmness. At higher acid concentrations, the gel formed was weak and fell apart during filtration after alcoholic precipitation. Therefore, since at a concentration of 5 g (100 g)$^{-1}$ of citric acid and an extraction time of 60 min, there was no significant difference in the average yields obtained when using higher concentrations of citric acid and longer times, it can be concluded that this is the most favourable combination for obtaining a better quality product.

4.2.2 Optimising the extraction of pectin from guava peel pulp

The average yields of pectin extracted from guava pulp and peel flour are shown in Table 8.

Table 8. Experimental design and average yield of pectin extracted from guava pulp and peel flour cv. *Pedro Sato*

Experiment	Coded variables		Original variables		Pectin yield [g (100 g)$^{-1}$]		
	Acid	Time	Acid [g (100 g)$^{-1}$]	Time (min)	Repetition 1	Repetition 2	Average [1]
1	-1	-1	3,5	35	7,76	6,87	7,31 ^{E}
2	1	-1	6,5	35	11,47	9,68	10,57 ^{B}
3	-1	1	3,5	85	9,10	8,64	8.87 ^{C,D,E}
4	1	1	6,5	85	10,16	11,12	10.64 ^{B,C,D}
5	-1,414	0	3	60	7,91	9,08	8.49 ^{D,E}
6	1,414	0	7	60	13,73	12,74	13,24 ^{A}
7	0	-1,414	5	25	5,59	6,24	5,91 ^{F}
8	0	1,414	5	95	12,83	12,87	12.85 ^{A,B}
9	0	0	5	60	11,30	11,52	11.41 ^{A,B}
10	0	0	5	60	11,94	11,69	11.81 ^{A,B}
11	0	0	5	60	11,37	11,36	11.36 ^{A,B,C}

1- Means followed by equal superscript letters in the column do not differ at the 5% probability level using the Tukey test.

The average yields observed ranged from 5.91-13.24 g (100 g)$^{-1}$, of which Experiments 6, 8, 9, 10 and 11 did not differ ($P > 0.05$) from each other. The highest pectin yield values were observed in these experiments, from 11.36 g (100 g)$^{-1}$ to 13.24 g (100 g)$^{-1}$. The average yields of experiments 2, 4, 8, 9, 10 and 11 also did not differ (P > 0.05) from each other, as did experiments 3, 4, and 11; 3 and 5; and 1 and 5. Experiment 7 had the lowest yield and differed (P < 0.05) from the others.

The analysis of variance and the effects of citric acid concentration and time on the yield of pectin extraction from guava peel pulp flour are shown in Table 9 and Figure 11.

Table 9. Analysis of variance for the adjusted model of pectin extraction yield from guava cv Pedro Sato pulp and peel flour as a function of citric acid concentration [g (100 g)$^{-1}$] and extraction time (min).

Factor analysed	Sum of squares	Degrees of freedom	Square medium	F	P
(1) Acid (L)	**34,15**	**1**	**34,15**	**25,68**	**0,0002**
Acid (Q)	3,46	1	3,46	2,60	0,1500
(2) Time (L)	**32,44**	**1**	**32,44**	**24,39**	**0,0003**
Time (Q)	**18,36**	**1**	**18,36**	**13,80**	**0,0030**
1L by 2L	1,H	1	1,H	0,83	0,4040
Lack of adjustment	**19,90**	**3**	**6,63**	**20,25**	**0,0001**
Total Waste	21,28	13	1,33		
	110,80	21			

Coefficients in bold are significant.

The results obtained from the analysis of variance show that the linear effects for citric acid concentration and extraction time and the quadratic effect of the time variable were significant (P < 0.05), but the acid (Q) and the interaction between the two variables (acid and time) were not significant (P > 0.05). The Pareto diagram (Figure 11) shows the influence of the positive linear effects of citric acid concentration and time on pectin yield, followed by the negative quadratic effect of extraction time. Fertonani (2006) found similar behaviour of the independent variables (acid concentration and extraction time) for the extraction of apple pomace pectin with nitric acid, in which the linear effects were positive and the quadratic effects negative. Kliemann (2006) in his study of the extraction of pectin from the peel of yellow passion fruit with various acids and considering the variables temperature, time and pH, observed negative linear effects for the pH of the solution and positive for the extraction temperature and negative quadratic effects for both variables in the extraction with citric acid; in the extraction with hydrochloric acid, the temperature and extraction time showed positive linear effects and negative quadratic effects, and with nitric acid the temperature and time showed only positive linear effects. Figure 12 shows the relationship between the standardised effects and the expected values using the complete mathematical model.

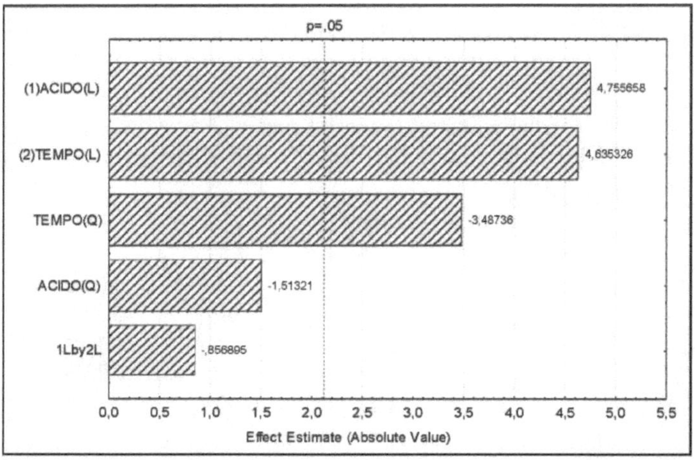

Figura 11. Effect of the variables "citric acid concentration" and "time" and their interactions on the yield of pectin extraction from guava pulp flour with Pedro Sato peel.

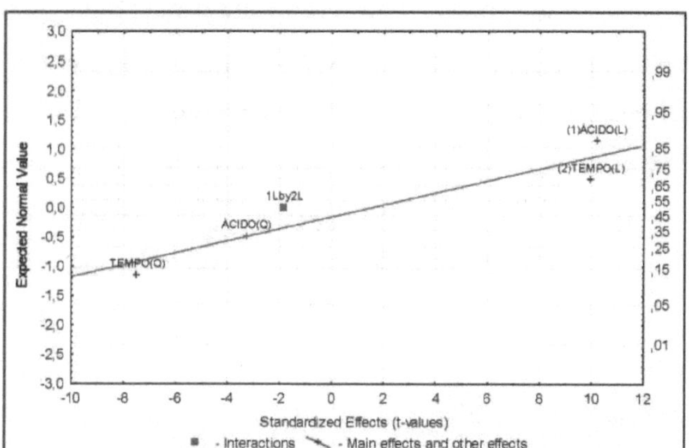

Figura 12. Effect of extracting pectin from guava pulp flour with Pedro Sato peel.

The mathematical model was adjusted by eliminating the non-significant factors: acid

(Q) and the interaction of acid (L) versus time (L), since the influence of these factors on the extraction yield was not observed in the analysis of variance and their lack of fit was significant (Table 9). The adjusted model proposed to describe the pectin extraction yield from guava peel pulp flour is shown in Equation 11.

$$Yield = 11.04 + 1.47 * Ac + 1.43 * t - 1.11 * t^2 \quad \text{Equation (11)}$$

Where:

Rend = pectin extraction yield [g (100 g)$^{-1}$];

Ac = citric acid concentration [g (100 g)$^{-1}$];

t = extraction time (min).

Table 10 shows the analysis of variance of the adjusted model.

Table 10. Analysis of variance for the adjusted model of pectin extraction yield from guava cv Pedro Sato pulp and peel flour as a function of citric acid concentration [g (100 g)$^{-1}$] and extraction time (min).

Factor analysed	Degrees of freedom	Sum of squares	Mean square	F	P
Model	3	82,3661	27,4554	17,44	0,0001
Error	18	28,3416	1,5745		
Corrected total	21	110,7077			

The adjusted model is significant (P=0.0001), and the coefficient of determination (R^2) obtained shows that 74.40 per cent of the variation in the response is explained by the model. The coefficient of variation (CV) for this analysis was 12.27 per cent.

Figures 13 and 14 show the response surfaces for the pectin extraction yield from guava peel pulp flour as a function of citric acid concentration and extraction time, obtained at 97 °C.

The level curves of the response surface presented in Figure 13 show the influence of acid concentration on the extraction yield, a behaviour similar to that observed for the extraction of pectin from guava pulp (item 4.2.1, Figure 9). However, the influence of the linear effects of acid concentration and time was close, whereas in the extraction of pectin from pulp, the influence of acid concentration was greater than that of time.

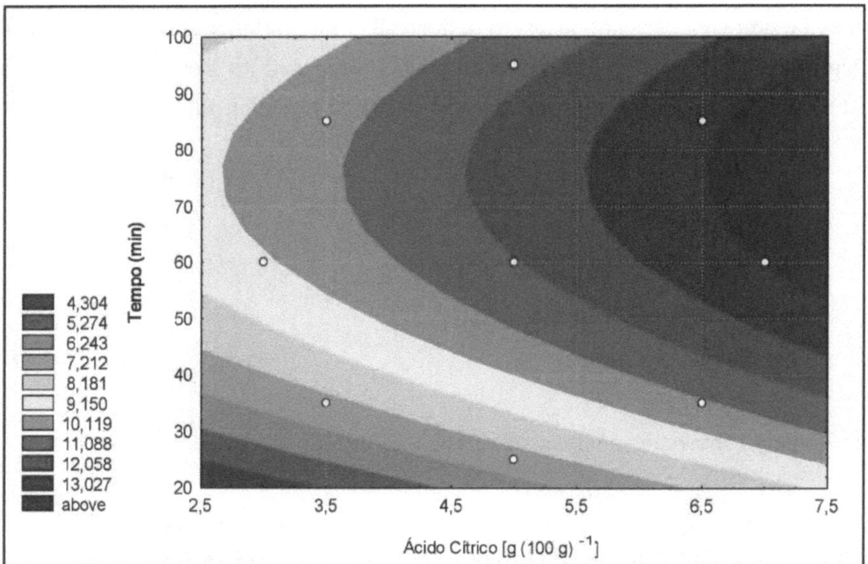

Figure 13. Contour plot of the yield [g (100 g)$^{-1}$] of pectin extraction (legend) from guava cv Pedro Sato pulp and peel flour as a function of time (min) and citric acid concentration [g (100 g)$^{-1}$].

According to the graph in Figure 13, the optimum region (with the highest yield) for the pectin extraction process from guava pulp with peel is in the 55-95 min time range and at citric acid concentrations above 6.5 g (100 g)$^{-1}$. In order to reduce acid consumption and the production of more polluting effluents, the ideal conditions for obtaining maximum extraction would be 6.5 g (100 g)$^{-1}$ and an operating time of 75 minutes. However, under high extraction temperatures, prolonged times can lead to pectin degradation (PAGÁN et al., 2001). In the tests carried out based on the planning, it was found that the concentration of 5 g (100 g)$^{-1}$ of citric acid and an extraction time of 60 min produced the best gel formation, as observed visually by its firmness. In extractions with higher concentrations of acid, the gels formed during alcoholic precipitation were weak and fell apart during filtration to obtain pectin. Therefore, as in this condition there was no significant difference in the average yields obtained with the use of higher concentrations of citric acid and longer times, it can be concluded that this is the most favourable combination for obtaining a better quality product.

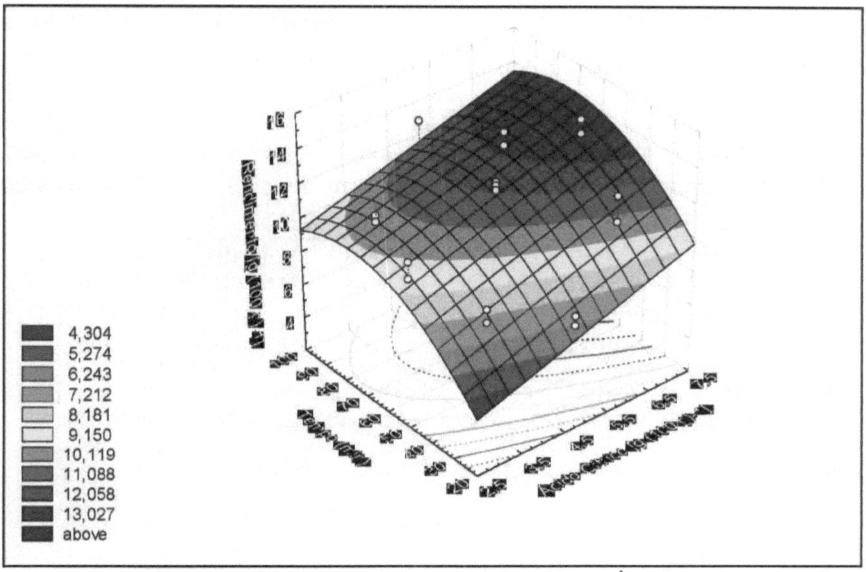

Figure 14. Response surface plot of the yield [g (100 g)$^{-1}$] of pectin extraction from guava cv Pedro Sato pulp flour as a function of time (min) and citric acid concentration [g (100 g)$^{-1}$].

Few studies have been found in the literature on the use of vegetable peel; Kliemann (2006) obtained a better yield of pectin under the conditions of extraction at 90 °C, pH 1.2 with citric acid for 10 min in samples with and without flavedo from yellow passion fruit peel. Kalapathy and Proctor (2001) studied how to obtain pectin from soya husks; the authors assessed the extraction yield using hydrochloric acid at concentrations of 0.06 to 0.18 g 100 g^{-1} (0.05 and 0.3 N) and found that the highest yields (17 and 19%) were obtained at concentrations of 0.06 and 0.12 g 100 g^{-1}, with no significant difference between the two concentrations.

4.3 TITRIMETRIC CHARACTERISATION OF PECTIN

The quality characteristics of the pectins assessed included the relative content of galacturonic acids (AUA) and methoxyls (MeO), the sum of which represents the percentage of polygalacturonic acid, the content of neutral sugars (the remaining percentage of the difference between 100 and the AUA and MeO contents) and the degree of esterification, which is independent of the amount of mass, since the formula used considers the ratio between the methoxylated AUA fraction and the total AUA fraction. The results of these parameters are shown in Table 11.

Table 11. Titratometric quality of pectins extracted from guava pulp and guava peel flours with citric acid at 5 [g (100 g)$^{-1}$], 60 min extraction time at 97 °C and commercial pectin.

Pectin	Galacturonic acids (AUA) [g (100 g)$^{-1}$]	Methoxyl (MeO) [g (100 g)$^{-1}$]	Acid fraction [g (100 g)$^{-1}$]	Neutral Sugars [g (100 g)$^{-1}$]	Degree of esterification (DE) [g (100 g)$^{-1}$]
Guava pulp flour	64,88 ^{C} ± 0,11	4,45 ^{B} ± 0,40	69,33 ^{C} ± 0,29	30,67 ^{A} ± 0,29	38,94 ^{B} ± 1,50
Guava pulp and peel flour	68,84 ^{B} ± 0,77	4,95 ^{B} ± 0,16	73,81 ^{B} ± 0,73	26,19 ^{B} ± 0,73	40,99 ^{B} ± 1,30
Commercial	75,38 ^{A} ± 0,54	9,56 ^{A} ± 0,12	84.94 ^{A} ± *0.41*	15.06 ^{C} ± *0.41*	72,01 ^{A} ± 1,27

Averages followed by equal superscript letters in the column do not differ at the 5% probability level using the Tukey test.

The conditions for gelling a pectin are determined by its degree of esterification, thus enabling its application. Pectin with a high degree of esterification gels under pH conditions of less than 3.6 and sucrose concentrations of between 60-65%, while pectin with a low degree of esterification gels in the absence of sugar and requires the presence of bivalent cations such as calcium, a substance commonly used in foodstuffs, and can gel in a pH range of 2.5 to 6.5 (CALLIARI, 2004).

The pectins obtained from guava pulp and guava peel flours were similar ($P > 0.05$) in terms of methoxyl content (MeO) and differed ($P < 0.05$) from commercial pectin (9.56%), with lower values (4.45 and 4.95%, respectively). However, the MeO values were similar (4.81%) to those found by Fertonani et al. (2006) for apple pomace pectin extracted with nitric acid at a concentration of 100 mMol (L)$^{-1}$ at 97 °C for 10 min.

With regard to AUA, although the values obtained showed a significant difference ($P < 0.05$), they are close to the minimum percentage of galacturonic acid (65%) for commercial pectins established by the FAO (FOOD AND AGRICULTURE ORGANISATION OF THE UNITED NATIONS, 2007). Values close to the AUA in guava pulp flour were found by Calliari (2004) in orange pomace pectin extracted with 4% pomace, 3.5% citric acid at 75 °C, with 64.10% being reported.

The pectins obtained in the laboratory showed low esterification (38.94% and 40.99%). Calliari (2004) obtained similar average values (37.50%) for pectin extracted with citric acid from orange pomace under the conditions of 4% pomace, 3.5% citric acid at 75 °C. Kliemann (2006) also extracted low esterification pectin (45.94%) using citric acid at pH 1.2 at 90 °C for 10 min and passion fruit peels as raw material. Marcon et al. (2005) extracted low esterification pectin (14.3% to 29.4%) from apple pomace using 5% citric acid, extraction times between 30-80 min and temperatures between 50-100 °C.

The pectins obtained from the flours studied in this study had a degree of esterification of less than 50%, and are characterised as low esterification pectins. They can form stable gels in the absence of sugar, but require the presence of bivalent ions. Compared to high-esterification pectins, they are less sensitive to pH changes and can form gels in the pH 2.6-6.0 range. They can also be used as a thickener and stabiliser in food emulsions (SHARMA; LIPTAY; LE MARGUER, 1998; YAPO et al. 2007). In recent years, pectin has also been used as a soluble dietary fibre due to its beneficial physiological effects on the human body, as it prevents the harmful effects of pathologies and has an influence on reducing levels of cholesterol, lipoproteins, bile acids and glucose, as well as promoting the complexation of heavy metals and their isotopes (CANTERI-SCHEMIN, 2003).

CHAPTER 5

CONCLUSION

The moisture content of the guava peel, pulp, pulp with peel and core flours proved to be suitable for storage at room temperature.

The guava pulp and guava peel flours had higher pectin contents than the guava peel and core flours, so they have potential for commercial extraction.

The process of extracting pectin with citric acid and alcoholic precipitation in a boiling system showed yields of over 11% for guava pulp and guava peel flour. For both, the optimum extraction conditions were: acid concentration of 5 g (100 g)$^{-1}$ and extraction time of 60 min.

The extracted pectins obtained under the optimum extraction conditions from the guava pulp and guava peel flours had a low degree of esterification with 38.94% and 40.99%, respectively. Despite the low degree of esterification, the pectins had a galacturonic acid concentration close to the commercial standard.

The extraction of pectin from guava pulp and guava peel flour is technically feasible, and these pectins can be used for gelling foods with a low sugar content and in the presence of a bivalent calcium ion in the formulation, as well as being used as a soluble dietary fibre, thickener and stabiliser.

REFERENCES

AZZOLINI, M. **Post-harvest physiology of "Pedro Sato" guavas: ripening stages and respiration pattern.**2002. 112f. Dissertation (Master of Science) - Luiz de Queiroz College of Agriculture, University of São Paulo, Piracicaba, 2002.

AZZOLINI, M.; JACOMINO, A. P.; SPOTO, M. H. F. Ripening stages and post-harvest quality of "Pedro Sato" guavas. **Revista Brasileira de Fruticultura,** Jaboticabal, v. 26, n. 1, p-29-31, 2004.

BATISTA, J. A.; TANADA-PALMU, P. S.; GROSSO, C. R. F. Effect of adding fatty acids to pectin-based films. **Ciência e Tecnologia de Alimentos,** Campinas, v. 25, n. 4, p. 781-788, 2005.

BERARDINI, N.; KNODLER, M.; SCHIEBER, A.; CARLE, R. Utilisation of mango peels as a source of pectin and polyphenolics. **Innovative Food Science &**

Emerging Technologies, Rotterdam, v. 6, n. 4, p. 442-452, 2005.

BRAZIL. Ordinance No. 398 of 30 April 1999. Approves the technical regulation that establishes the basic guidelines for analysing and proving functional and/or health properties claimed on food labelling. **Official Gazette of the Federative Republic of Brazil,** Brasília, DF, 30 April 1999. Available at: <http://www.anvisa.gov.br>. Accessed on: 10 Jan. 2008.

BRAZIL. Resolution No. 263 of 22 September 2005. Approves technical regulations for cereal products, starches, flours and meals. **Official Gazette of the Federative Republic of Brazil,** Brasília, DF, 22 September 2005. Available at: <http://www.anvisa.gov.br>. Accessed on: 10 July 2007.

BRUNINE, M. A.; OLIVEIRA, A. L.; VARANDA, D. B. Evaluation of the quality of **"Paluma" guava** pulp at -20°C. **Revista Brasileira de Fruticultura**, Jaboticabal, v. 25, n. 3, p. 394-396, 2003.

CALLIARI, C. M. **Thermal, chemical and enzymatic extraction of pectin from orange pomace.** 2004. 96 f. Dissertation (Master's Degree in Food Science) - Department of Food Technology, State University of Londrina. Londrina, 2004.

CANTERI-SCHEMIN, M. H. **Obtaining food-grade pectin from apple pomace.** 2003. 83 f. Dissertation (Master's Degree in Food Technology) - Technology Sector, Federal University of Paraná. Curitiba, 2003.

CANTERI-SCHEMIN, M. H.; FERTONANI, H. C. R.; WASZCZYNSKYJ, N.; WOSIACKI, G. Extraction of pectin from apple pomace. **Brazilian Archives of Biology and Technology**, Curitiba, v. 48, n. 2, p. 259-266, 2005.

CERQUEIRA, T. S. **Edible coatings on "Kumagai" guavas.** 2007. 70 f. Dissertation (Master's in Food Sciences) - Luiz de Queiroz College of Agriculture, University of São Paulo, Piracicaba, 2007.

CHISTÉ, R. C.; COHEN, K. O.; MATHIAS, E. A.; RAMOA JÚNIOR, A. G. A. Quality of dried cassava flour. **Ciência e Tecnologia de Alimentos**, Campinas, v. 26, n. 4, p. 861-864, 2006.

CHISTÉ, R. C.; COHEN, K. O.; MATHIAS, E. A.; RAMOA JUNIOR, A. G. A. Study of physicochemical and microbiological properties in the processing of manioc flour from the **d'água group. Ciência e Tecnologia de Alimentos**, Campinas, v. 27, n. 2, p. 265-269, 2007.

DA-SILVA, R.; FRANCO, C. M. L.; GOMES, E. Pectinases, hemicellulases and cellulases, action, production and application in food processing: a review. **Bulletin of the Brazilian Society of Food Science and Technology,** Campinas, v. 31, n. 2, p. 249-260, 1997.

DZIEZAK, J. D. A focus on gums: building in textures, stability, and emulsification are just some of ways in which gums aid products development. **Food Technology,** Chicago, v. 45, n. 3, p. 116-132, 1991.

FASOLIN, L. H.; ALMEIDA, G. C.; CASTANHO, P. S.; NETTO-OLIVEIRA, E. R.
Biscuits produced with banana flour: chemical, physical and sensory evaluations. **Ciência e Tecnologia de Alimentos**, Campinas, v. 27, n. 3, p. 524-529, 2007.

FERTONANI, H. C. R. **Establishment of a model for the acid extraction of pectin from apple pomace.** 2006. 82 f. Dissertation (Master's Degree in Food Science and Technology) - Department of Food Science and Technology, State University of Ponta Grossa, Ponta Grossa, 2006.

FERTONANI, H. C. R.; SCABIO, A.; CANTERI-SCHEMIN, M. H.; CARNEIRO, E. B. B.;
NOGUEIRA, A.; WOSIACKI, G. Influence of acid concentration on the extraction process and quality of apple pomace pectin. **Semina:** Ciências Agrárias, Londrina, v. 27, n. 4, p. 599-612, 2006.

FISHMAN, M. L.; CHAN, H. K.; HOAGLAND, P. D.; HOTCHKISS, A. T. Microwave- assisted extraction of lime pectin. **Food Hydrocolloids,** Oxford. v. 20, n. 8, p. 1170-1177, 2006.

FIETZ, V. R.; SALGADO, J. M.; Effect of pectin and cellulose on serum cholesterol and triglyceride levels in hyperlipidaemic rats. **Ciência e Tecnologia de Alimentos,** Campinas, v. 19, n. 3, p. 318-321, 1999.

FOOD AND AGRICULTURE ORGANISATION OF THE UNITED NATIONS. **Health (emulsifiers, stabilisers, thickeners and gelling agents in food) regulations 1994** (FAO). Available at:<http://www.fao.org.com>. Accessed on: 15.12.2007.

GIOSAFATTO, C. V. L.; MARINIELLO, L.; RING, S. Extraction and characterisation of Foeniculum vulgare pectins and their use for preparing biopolymer films in the presence of phaseolin protein. **Journal of Agricultural and Food Chemistry**, Easton, v. 55, n. 4, p. 1237-1240, 2007.

GNANASAMBANDAM, G.; PROCTOR, A. Preparation of soy hull pectin. **Food Chemistry,** London, v. 65, n. 4, p. 461-467, 1999.

GUILLOTIN, S. E.; BAKX, E. J.; BOULENGUER, P.; MAZOYER, J.; SCHOLS, H. A.; VORAGEN, A. G. J. Populations having different GalA block characteristics are present in commercial pectins which are chemically similar but have different functionalities.
Carbohydrate Polymers, Barking, v. 60, n. 3, p. 391-398, 2005.

GUTKOSKI, L. C.; NODARI, M. L.; JACOBSEN-NETO, R. Evaluation of wheat flours grown in Rio Grande do Sul in the production of biscuits. **Ciência e Tecnologia de Alimentos**, Campinas, v. 23, supl, p. 91-97, 2003.

HAMINIUK, C. W. I. **Study of the rheological and colorimetric behaviour of ternary mixtures and pectic systems of strawberry, blackberry and raspberry pulps.** 2007. *147f.* Thesis (Doctorate in Food Technology) - Technology Sector, Federal University of Paraná, Curitiba, 2007.

HANSSON, A.; ANDERSSON, J.; LEUFVEN, A. The effect of sugars and pectin on flavor release from a soft drink-related model system. **Food Chemistry**, London, v. 72, n. 3, p. 363368, 2001.

HOJO, R. H.; CHALFUN, N. N. J.; HOJO, E. T. D.; VEIGA, R. D.; PAGLIS, C. M.; LIMA, L. C. O. Production and fruit quality of 'Pedro Sato' guava trees subjected to different pruning periods. **Pesquisa Agropecuária Brasileira**, Brasília, v. 42, n. 3, p. 357-362, 2007.

ADOLFO LUTZ INSTITUTE. **Analytical standards of the Adolfo Lutz Institute.** São Paulo: IAL, 2005. 533p.

JACOMINO, A. P.; OJEDA, R. M.; KLUGE, R. A.; SCARPARE FILHO, J. A. Preservation of guavas treated with carnauba wax emulsions. **Revista Brasileira de Fruticultura**, Jaboticabal, v. 25, n. 3, p. 401-105, 2003.

JAY, J. M. **Microbiologia de Alimentos.** São Paulo: Artmed, 2005. 712p.

KALAPATHY, U; PROCTOR, A. Effect of acid extraction and alcohol precipitation conditions on the yeld and purity of soy hull pectin. **Food Chemistry,** London, v. 73, n. 4, p. 393-396, 2001.

KLIEMANN, E. **Extraction and characterisation of pectin from the peel of yellow passion fruit** (Passiflora edulis **flavicarpa).** 2006. 75 f. Dissertation (Master's in Food Science) - Federal University of Santa Catarina, Federal University of Santa Catarina. Florianópolis, 2006.

KRATCHANOVA, M.; PAVLOVAA, E.; PANCHEV, I. The effect of microwave heating of fresh orange peels on the fruit tissue and quality of extracted pectin. **Carbohydrate Polymers,** Barking, v. 56, n. 2, p. 181-185, 2004.

LAURENT, M. A.; BOULENGUER, P. Stabilisation mechanism of acid dairy drinks (ADD) induced by pectin. **Food Hydrocolloids,** Oxford, v. 17, n. 4, p.445-454, 2003.

LEVIGNE, S.; THOMAS, M.; RALET, M. C.; QUEMENER, B.; THIBAULT, J. F. Determination of the degrees of methylation and acetylation of pectins using a C18 columm and internal standards. **Food Hydrocolloids,** Oxford, v. 16, n. 6, p. 547-550, 2002.

LINHARES. L. A.; SANTOS, C. D.; ABREU, C. M. P.; CORRÊA, A. D. Chemical, physical and enzymatic transformations of 'PEDRO SATO' guavas treated post-harvest with calcium chloride and 1-methylcyclopropene and stored under refrigeration. **Ciência e Agrotecnologia**, Lavras, v. 31, n. 3, p. 829-841, 2007.

LIU, Y.; SHI, J.; LANGRISH, T. A. G. Water-based extraction of pectin from flavedo and albedo of orange peels. **Chemical Engineering Journal,** Lousane, v. 120, n. 3, p. 203-209, 2006.

MARCON, M. V.; VRIESMANN, L. C.; WOSIACKI, G.; BELESKI-CARNEIRO, E. Pectins from apple pomace. **Polímeros:** Ciência e Tecnologia, São Carlos, v. 15, n. 2, p. 127129, 2005.

MATTIUZ, B. H.; DURIGAN, J. F. Effect of mechanical injuries on **respiratory** processing **and chemical parameters of "Paluma" and "Pedro Sato" guavas. Revista Brasileira de Fruticultura,** Jaboticabal, v. 23, n. 2, p-29-31, 2001.

MAY, C. D. Industrial pectins: sources, production and applications. **Carbohydrate Polymers,** Barking, v. 12, n. 15, p. 79-99, 1990.

MESBAHI, G.; JAMALIAN, J.; FARAHNAKY, A. A comparative study on functional properties of beet and citrus pectins in food systems. **Food Hydrocolloids,** Oxford, v. 19, n. 4, p. 731-738, 2005.

MIRANDA, M. Z.; EL-DASH, A. Wholemeal flour from sprouted wheat. **Ciência e Tecnologia de Alimentos,** Campinas, v. 22, n. 3, p. 216-223, 2002.

NACHTIGALL, A. M.; SOUZA, E. L.; MALGARIM, M. B.; ZAMBIAZI, R. C. Light blackberry jam. **Bulletin of the Food Research and Processing Centre,** Curitiba, v. 22, n. 2, p. 337-354, 2004.

NAKAMURA, A.; YOSHIDA, R.; MAEDA, H. CORREDIG, M. The stabilising behaviour of soybean soluble polysaccharide and pectin in acidified milk beverages. **International Dairy Journal,** Barking. v. 16, n. 4, p. 361-369, 2006.

ORDONEZ-PEREDA, J. A. Carbohydrates. In:. **Food technology:** components of foods and processes. São Paulo: Artmed, v. 1, 2005. chap. 4, p. 63-79.

PAGÁN, J.; IBARZ, A. Extraction and rheological properties of pectin from fresh peach pomace. **Journal of Food Engineering,** Essex, v. 39, n. 2, p. 193-201, 1999.

PAGÁN, J.; IBARZ, A.; LLORCA, M.; PAGÁN, A.; BARBOSA-CÁNOAS, G. V. Extraction and chacarterisation of pectin from stored peach pomace. **Food Research International,** Barking, v. 34, n. 7, p. 605-612, 2001.

PIEDADE, J.; CANNIATTI-BRAZACA, S. G. Comparison between the effect of pineapple residue (stems and leaves) and high methoxylation citrus pectin on blood cholesterol levels in rats. **Cência e Tecnologia de Alimentos,** Campinas, v. 23, n. 2, p. 149-156, 2003.

RANGANNA, S. **Manual of analysis of fruit and vegetable products.** New Delhi: Tata McGraw Hill Publishing Company Limited, 1977. 634 p.

RIBEIRO, E. P.; SERAVALLI, E. A. G. **Química de Alimentos.** São Paulo: Edgard Blucher, 2004. 184p.

RODRIGUES, M. A.; IEMMA, A. F. **Planning experiments and optimisation:** a sequential planning strategy. Campinas: Casa do Pão, 2005. 326 p.

SANJINEZ-ARGANDONA, E. J. **Osmotically dehydrated and dried guavas:** evaluation of a semi-continuous osmotic system, drying and quality. 2005. 172 f. Thesis (Doctorate in Food Engineering) - State University of Campinas, Faculty of Food Engineering, Campinas, 2005.

SATO, A. C. K.; SANJINEZ-ARGANDONA, E. J.; CUNHA, R. L. Evaluation of the physical, chemical and sensory properties of industrialised guavas in syrup. **Ciência e Tecnologia de Alimentos,** Campinas, v. 24, n. 4, p. 550-555, 2004.

SHARMA, S. K.; LIPTAY, A.; LE MARGUER, M. Molecular characterisation, physico- chemical and functional properties tomato fruit pectin. **Food Research International**, Barking, v. 30, n. 7, p. 543-547, 1998.

SINGTHONG, J.; NINGSANOND, S.; CUI, S. W.; GOFF, D. Extraction and physicochemical characterisation of Krueo Ma Noy pectin. **Food Hydrocolloids,** Oxford, v. 19, n. 5, p. 793-801, 2005.

STATISTICA for Windows-Release 5.0. Tulsa (USA): StatSoft Inc., 1995.

SURH, J.; DECKER, E. A.; MC-CLEMENTS, D. J. Influence of pH and pectin type on properties and stability of sodium-caseinate stabilised oil-in-water emulsions. **Food Hydrocolloids,** Oxford, v. 20, n. 5, p. 607-618, 2006.

TERPSTRA, A. H. M; LAPRE, J. A; DE VRIES, H. T.; BEYNEM, A. C. Dietary pectin with high viscosity lowers plasma and liver cholesterol concentration and plasma cholesteryl ester protein activity in hamsters. **The Journal of Nutrition,** Rotterdam, v. 128, n. 11, p. 19441949, 1998.

THAKUR, B. R.; SINGH, R. K.; HANDA, A. K. Chemistry and uses of pectin - a review. **Critical Reviews in Food Science and Nutrition,** Boca Raton, v. 37, n. 1, p. 47-73, 1997.

THE SAS system for Windows 3.95-Release 6.08. Cary (USA): SAS Institute, Inc., 1992.

VILA, M. T. R; LIMA, L. C. O.; VILAS-BOAS, E. V. B.; HOJO, E. T. D.; RODRIGUES, L. J.; PAULA, N. R. F. Chemical and biochemical characterisation of guavas stored under refrigeration and modified atmosphere. **Ciência e Agrotecnologia**, Lavras, v. 31, n. 5, p. 14351442, 2007.

WILLATS, W. G. T.; KNOX, J. P.; MIKKELSEN, J. D. Pectin: new insights into and old polymers are starting to gel. **Trends in Food Science & Technology**, Cambridge, v. 17, n. 3, p. 97-104, 2006.

XISTO, A. L. R. P.; ABREU, C. M. P.; CORRÊA, A. D.; SANTOS, C. D. Texture of "Pedro Sato" guavas submitted to calcium chloride application. **Ciência e Agrotecnologia**, Lavras, v. 28, n. 1, p. 113-118, 2004.

YAPO, B. M.; ROBERT, C.; ETIENNE, I.; WATHELET, B.; PAQUOT, M. Effect of extraction conditions on the yield, purity and surface properties of sugar beet pulp pectin extracts. **Food Chemistry**, London, v. 100, n. 4, p. 1356-1364, 2007.

ZHONGDONG, L.; GUOHUA, W.; YUNCHANG, G.; KENNEDY, J. F. Image study of pectin extraction from orange skin assisted by microwave. **Carbohydrate Polymers**, Barking, v. 64, n. 4, p. 548-552, 2006.

FSC
www.fsc.org
MIX
Papier | Fördert
gute Waldnutzung
FSC® C083411

Kolibri 360 GmbH
Ferdinand-Jühlke-Straße 7
99095 Erfurt, Deutschland
produktsicherheit@kolibri360.de

Druck:
CPI Druckdienstleistungen GmbH
im Auftrag der
Kolibri 360 GmbH
Ferdinand-Jühlke-Str. 7
99095 Erfurt